FIRST BOOK OF STARS

SIR PATRICK MOORE

AM

First published 2010

Amberley Publishing
Cirencester Road, Chalford,
Stroud, Gloucestershire, GL6 8PE

www.amberley-books.com

British Library Cataloguing in Publication Data.
A catalogue record for this book is available from the British Library.

ISBN 978 1 84868 291 7

Typesetting and origination by Amberley Publishing
Printed in Great Britain

Contents

Foreword by Brian May

Sir Patrick Moore is the quintessential icon of astronomy on TV, and a British National Treasure. But to astronomers everywhere, both amateur and professional, he is even more than this. He is the very reason that they became astronomers. For fifty years he has been bringing the wonders of the Universe into our living rooms, turning our heads towards the night sky, igniting our curiosity through his own passion for all things astronomical.

For me, as a small child, in the 1950s, it was magical to be allowed to stay up late and watch *The Sky at Night* — in black and white, of course, because television was young — there was only one station — and see Patrick Moore open doors to untold mysteries — stars, planets, comets, nebulae, galaxies, eclipses, and a sense of the unknowable infinite. One of Patrick's most famous phrases has always been the conclusion, 'we just don't know' ... signalling the truth — that the subject is always full of more questions than answers, our knowledge evolving almost minute by minute.

All of us were inspired, as youngsters, to take up the challenge offered by Patrick, and many of us followed the dream to actually become astronomers ourselves.

Now, with this lovely book, beautifully illustrated with the newest astrophotographs, Patrick appeals to yet another generation of potential young astrophysicists, astrobiologists, and cosmologists. There is no-one better qualified than Patrick to channel the latest knowledge of the Universe to young people. Looking up at the stars is where it all begins, of course, but this book takes us beyond the night sky, out into Space, bringing the reader up close to all the major phenomena of the known Universe. Sir Patrick speaks in simple terms, but never underestimates his readers — he knows that young minds are sharp and clear, and uncluttered with preconceptions. For young astronomers of all ages, *First Book of Stars* is destined to become a new classic.

Brian May, 2010

Acknowledgements

Before this book went to print, it was read through by several critics, all aged seven and under. In particular Cameron MacPherson and Charles Galloway made some very useful suggestions for improvements; accordingly I amended the text, and all their recommendations have been followed. My thanks too to William Galloway, Hannah MacPherson, Patrick Wilson and Annalieze Lambert. Particular thanks to the over sevens Trevor Little, Trudie Rayner and Bruce Kingsley. I hope I haven't left anybody out! I hope that this little book will help some of those seven-year-olds who are just starting to take an interest in the sky.

Sir Patrick Moore, 2010.

A Note for Helpers

In this book I have done my best to keep to words that a seven-year-old can read. Some words, which could cause problems, are listed below, and it might be a good idea to go through these before starting on the text. About the star names I can do nothing, but one soon becomes used to them!

Some Key Words

mountains
binoculars
Solar System
telescopes
meteor
meteorite
eclipse
constellation
diagram
Galaxy
appearance
nebula
Universe
radio

CHAPTER ONE

The Sky Above Us

Are you interested in the stars? If so, I will try to help you make a start. Look up in the daytime, when there are no clouds. The sky is blue, and the Sun is very bright. Do not look the Sun for long, or you will hurt your eyes. At night you will see the stars, and for many nights each month you will see the Moon too. You cannot see stars in the daytime, because the sky is too bright.

The Sun is a star, and seems so bright and hot only because it is much closer than the other stars. If it were just as far away, it would look like one of the stars shown in the picture opposite.

The Sun is much larger than the Earth. It is not solid and rocky. Like all stars it is made up of hot gases. Some of the stars you see at night are really much bigger and hotter than the Sun.

Cameron MacPherson using binoculars to observe the sky. Remember never to look at the Sun!
Trudie Rayner

A telescope used to look up at the sky.

North, South, East and West. This is the weather vane on top of my house!

CHAPTER TWO

The Earth and the Moon

The Earth is what we call a 'planet'. It moves round the Sun, and takes one year (365 days) to go once round. That is why your birthday comes only once a year!

The Earth spins round once in twenty-four hours. You can show this by putting a stick through a ball of wool and spinning it. The Sun sends us our light, but it can shine on only half of the Earth at a time. This is why we have day and night.

Opposite: Earth as seen from space. *NASA*

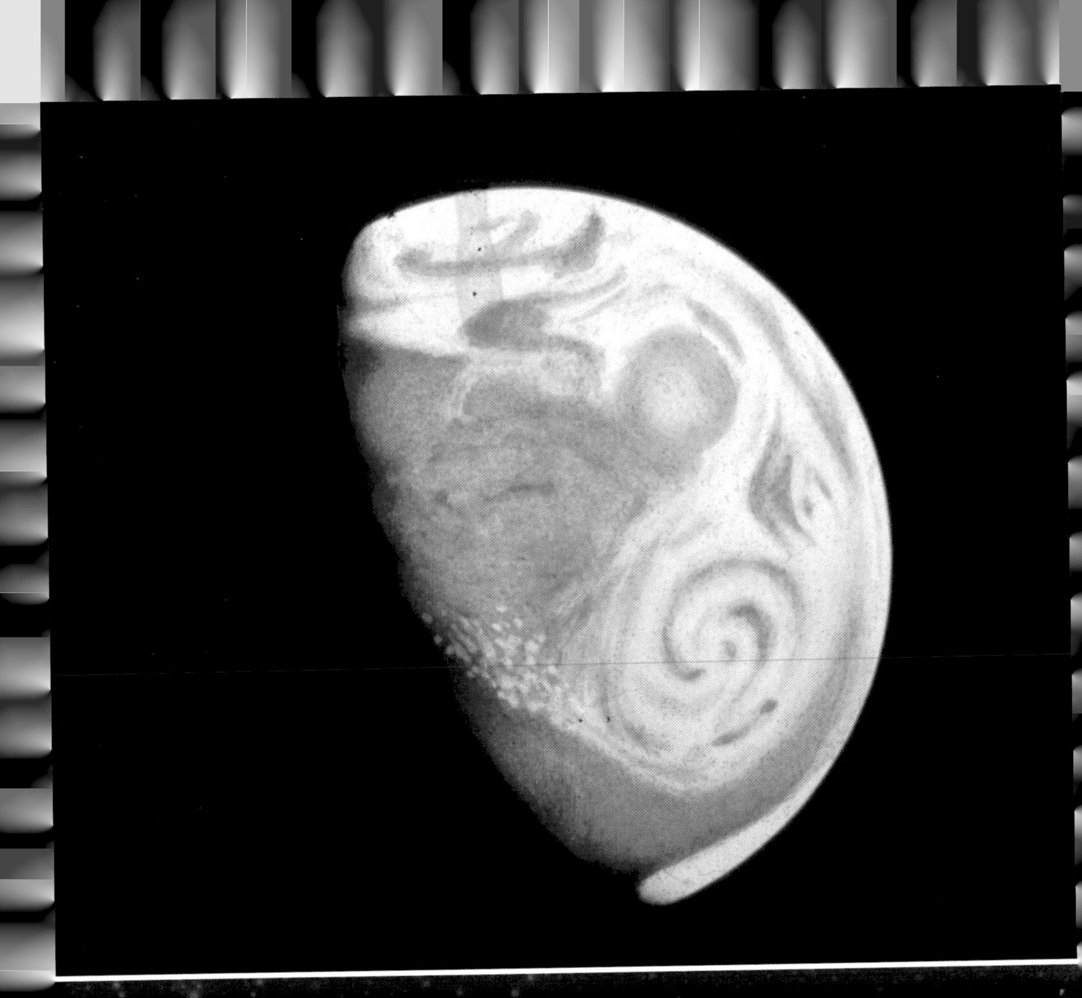

The Earth from space. *Paul Doherty.*

At night you often see the Moon. It has no light of its own, and shines only because it is lit up by the Sun. It moves round the Earth, and takes four weeks to go round once.

Spacemen working with rockets above the Earth. *Paul Doherty.*

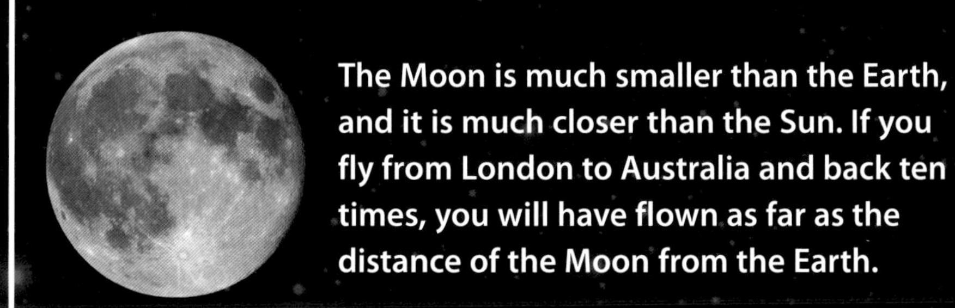

The Moon is much smaller than the Earth, and it is much closer than the Sun. If you fly from London to Australia and back ten times, you will have flown as far as the distance of the Moon from the Earth.

CHAPTER THREE

On the Moon

There is no air on the Moon, so that you could not live there. Without air you cannot have water. On the Moon we can see what we call seas, but there has never been any water in them. With no air and no water, there can be no life on the Moon.

Rockets have been sent to the Moon and men have been there. The first men to land on the Moon did so in 1969. No men have been there since 1972.

Opposite: A future base on the Moon. The people there would be living in domes. *Paul Doherty.*

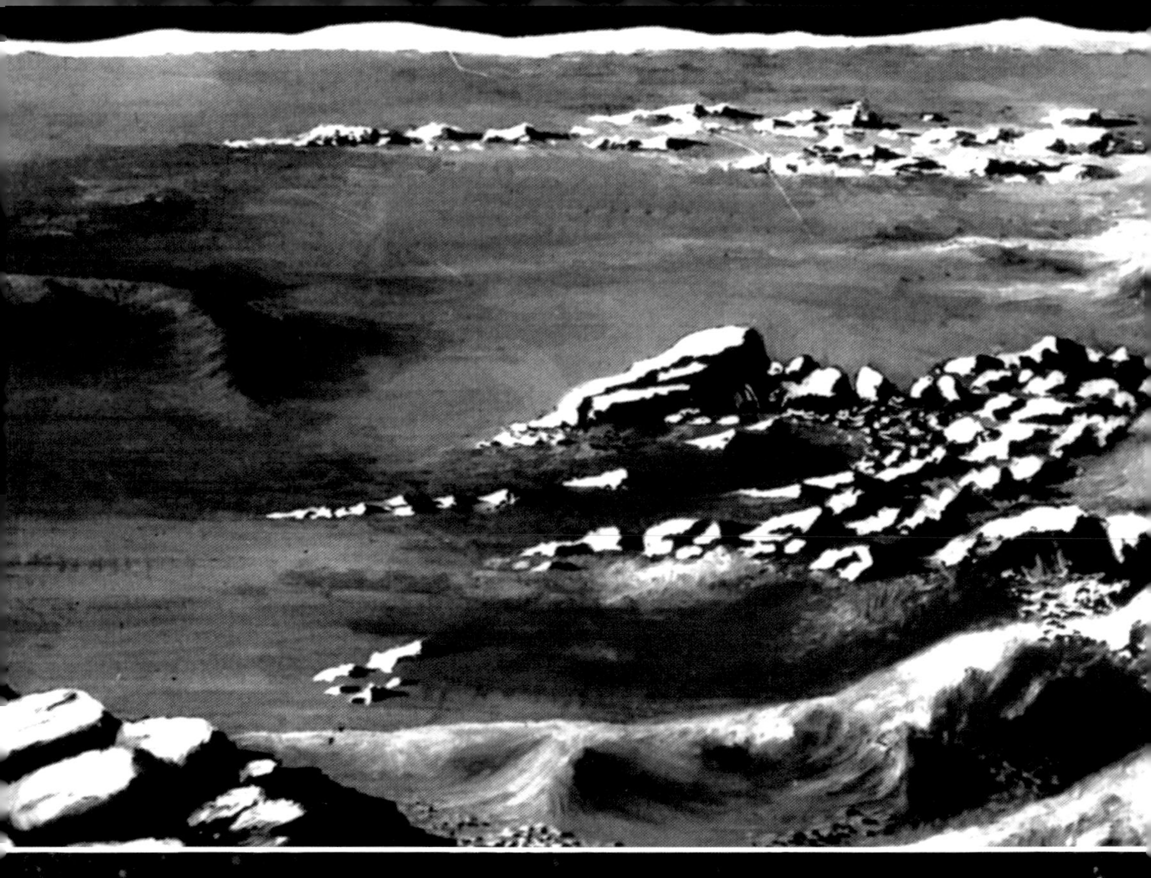

On the surface of the Moon, a scene of lava and mountains. *Paul Doherty.*

On the Moon you would feel much less heavy than you do at home. The sky would be black even in the daytime. A day on the Moon is fourteen times as long as ours.

The Sun can light up only half of the Moon. When the unlit side of the Moon is turned toward us, we cannot see the Moon. We call this a New Moon. If all the sunlit side is turned toward us the Moon is full.

At other times we may see only part of the sunlit side. The Moon seems to change shape from night to night. After a New Moon we see half of the sunlit side.

A crater on the Moon called Plato. It was made by a large body smashing nto the Moon and is sixty miles across so all of London could be placed

Hadley Rille mountains and craters on the Moon. *Trevor Little.*

Mountains and Craters on the Moon

There are mountains on the Moon, just as there are on Earth, and in the sunlight you can see their shadows. Some of the mountains are very high. On the Moon we see craters. You could say that a crater is a hole with a wall round it! Some of the craters are so big that you could put all of London inside them.

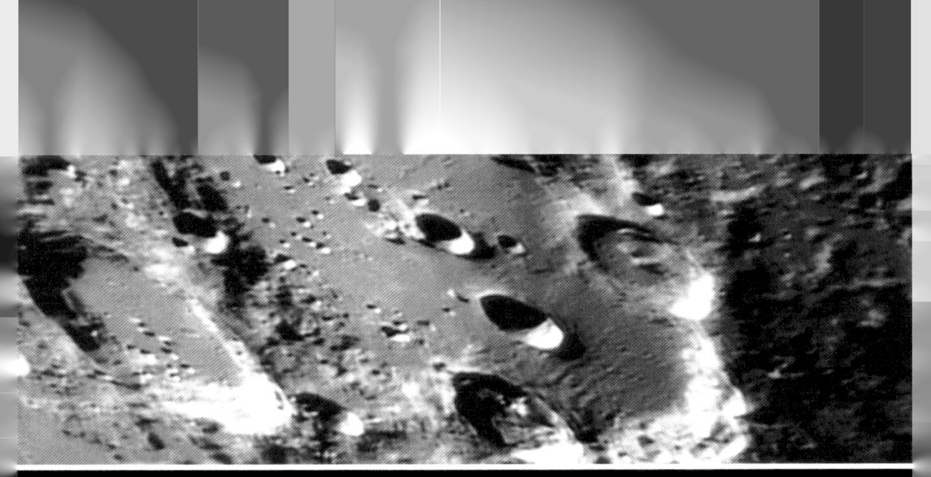

A huge crater on the Moon. It's nearly 150 miles across! *Bruce Kingsley.*

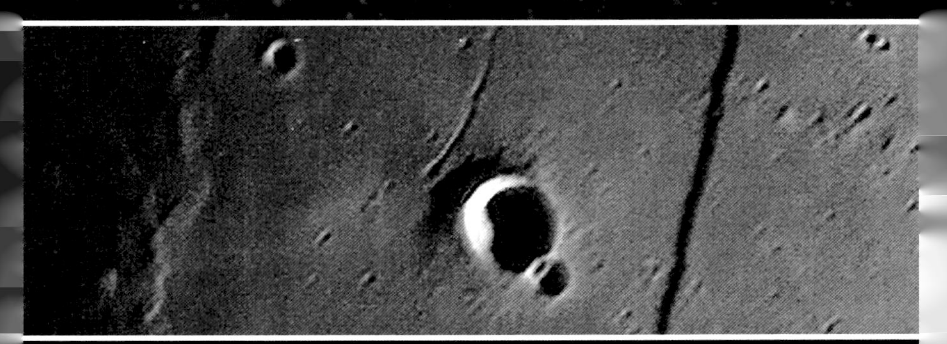

The streight wall, which is not a streight or a wall. The ground to the right is low down, so the high ground casts a shadow. *Trevor Little.*

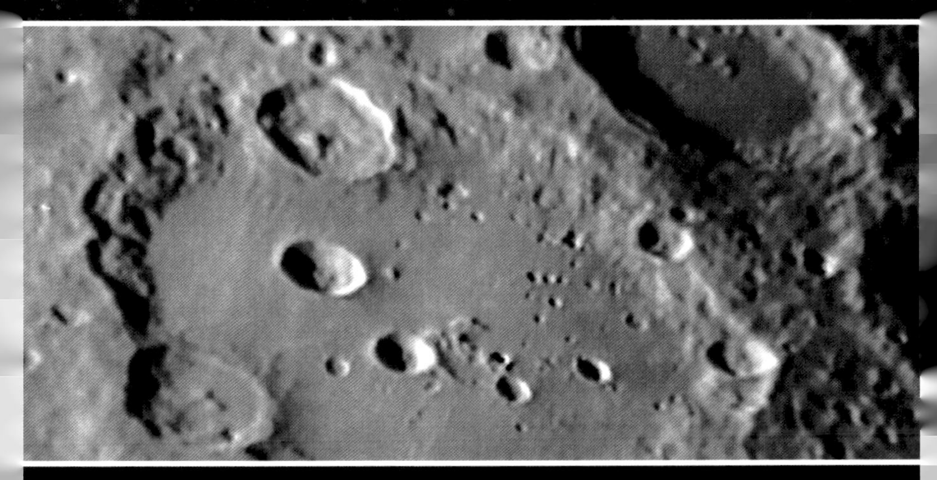

The crater Clavius, which is nearly 100 miles across with a chain of craters on its floor. *Trevor Little.*

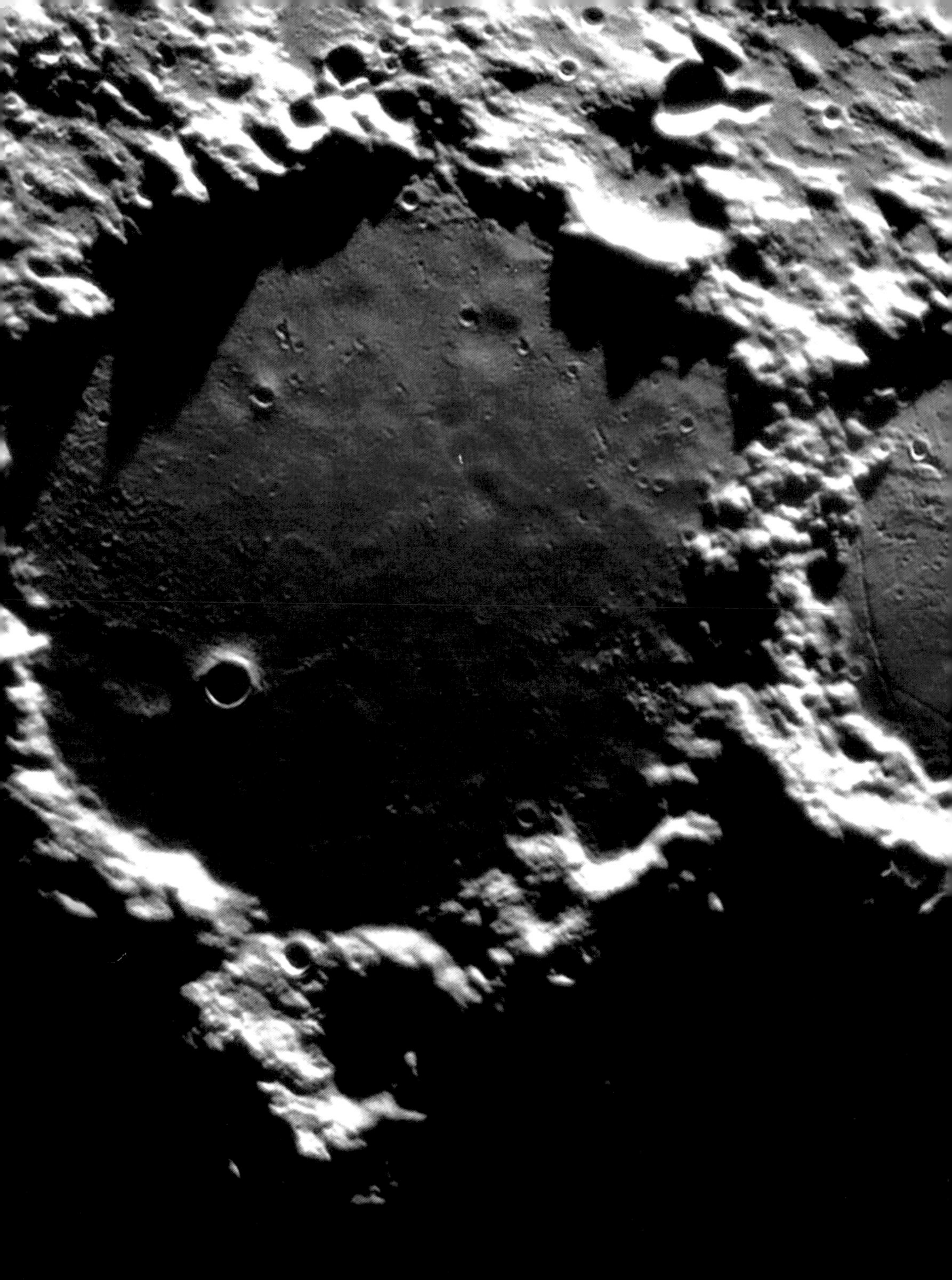

Three large craters on the Moon. The bottom one is 100 miles across and has a high central peak in the middle of it. *Nick Smith.*

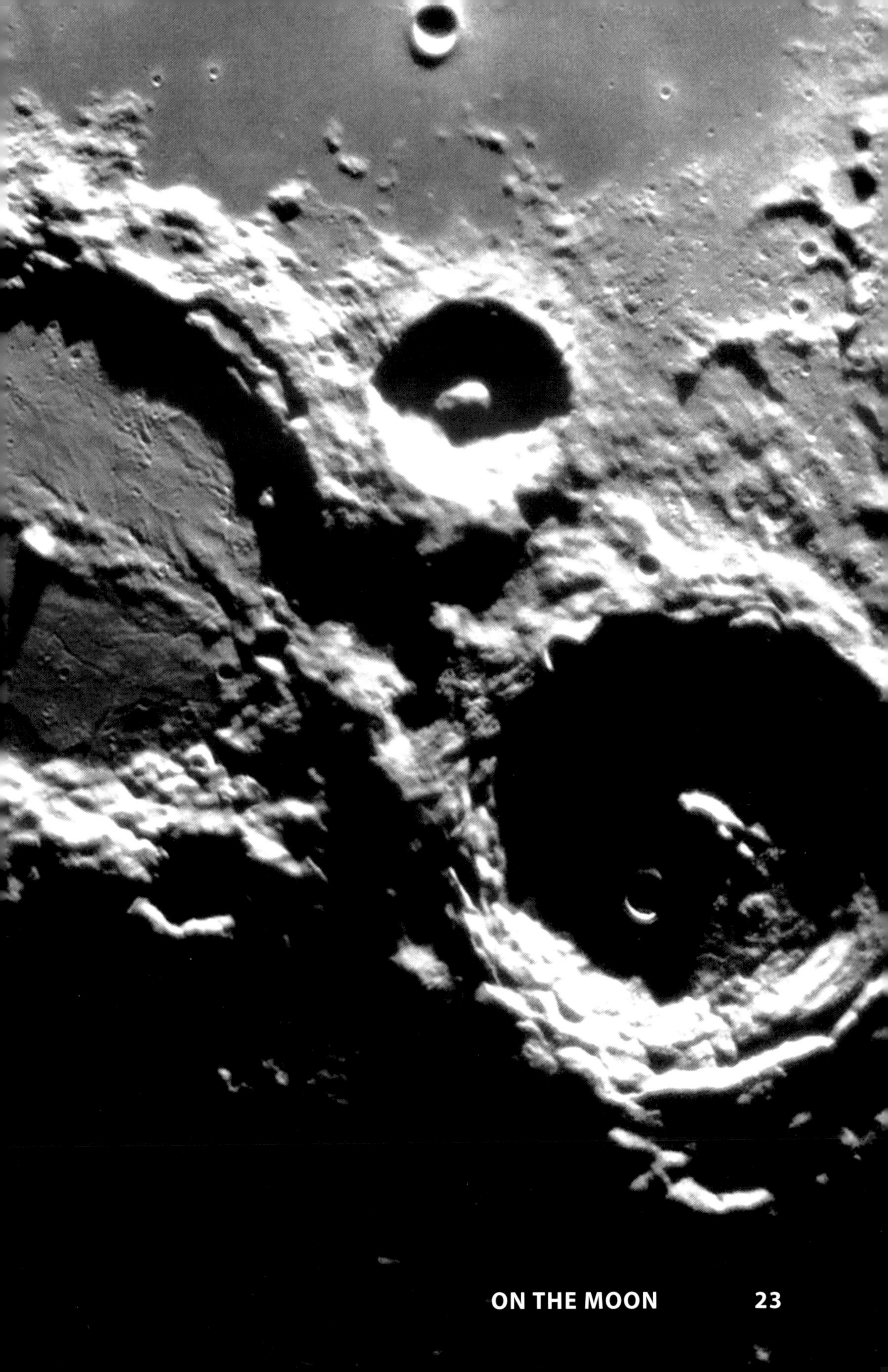

CHAPTER FOUR

Telescopes

To see the Moon and planets well, you will need a telescope. A telescope uses pieces of glass, called lenses, to make far-away things seem close. You can also use binoculars, which you can hold in your hand. A pair of binoculars is really two small telescopes joined together.

At the seaside, you can use a telescope to look at ships. In the sky, a telescope will show you the mountains and craters of the Moon.

Opposite: 3-inch refractor. This means that the lens at the top of the tube collecting the light is 3 inches across.

Above: Cameron MacPherson looking at the sky using my telescope.

Left: A 5-inch refracting telescope — the glass lens in the telescope is 5 inches across.

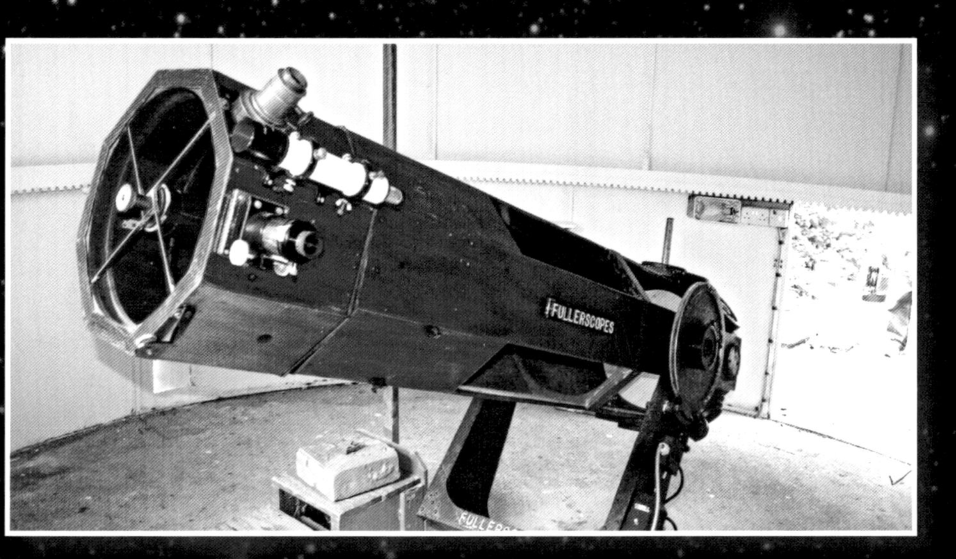

Above: The big telescope in my observatory. The light is collected by a mirror 15 inches across.

Below: Me standing by my main telescope in my observatory.

Above: The telescope at Jodrell Bank in Cheshire which collects radio waves from the sky. *Patrick Moore.*

Left: Josh Neville looks through my largest telescope with a 15-inch mirror. *Trevor Little.*

Charles Galloway uses a telescope pointing at the Sun to send a picture of the Sun onto a white card. *Trudie Raynor.*

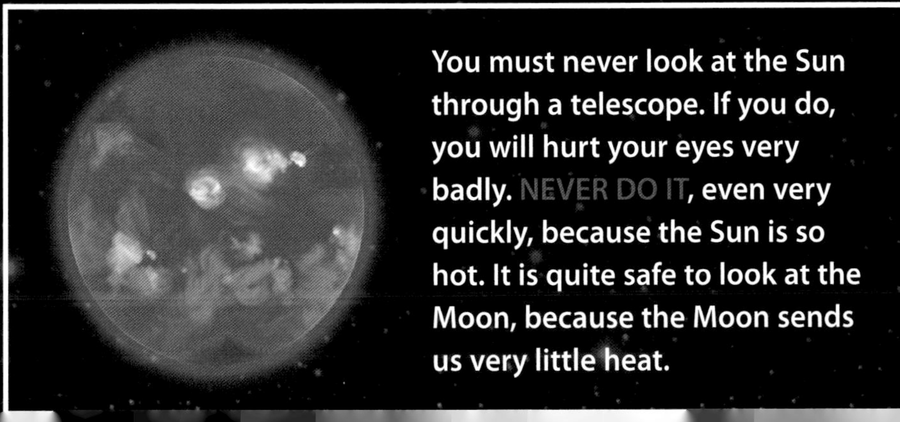

You must never look at the Sun through a telescope. If you do, you will hurt your eyes very badly. NEVER DO IT, even very quickly, because the Sun is so hot. It is quite safe to look at the Moon, because the Moon sends us very little heat.

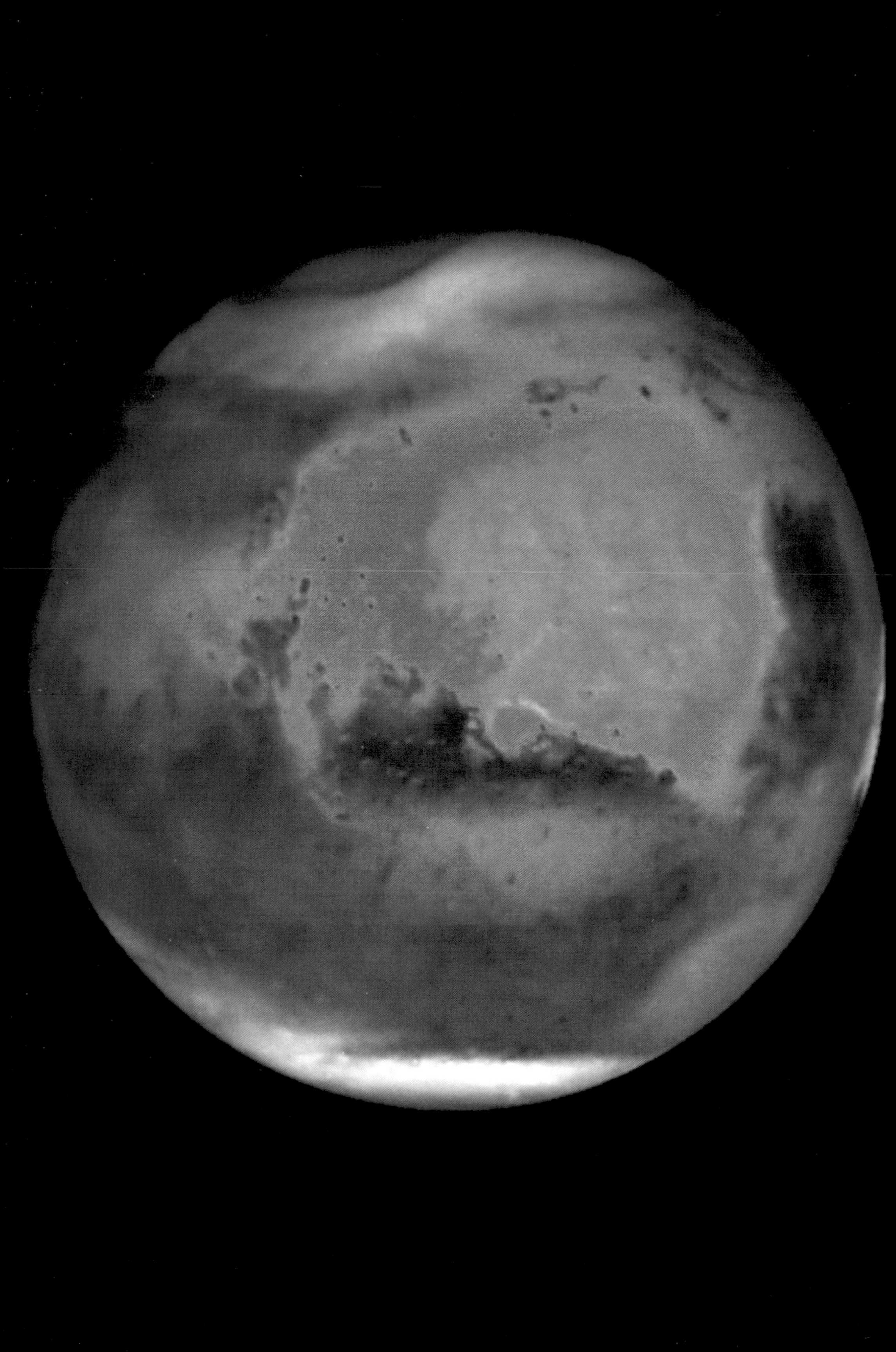

CHAPTER FIVE

The Planets

The Earth is a planet moving round he Sun. It is not the only planet. There are seven others, some bigger than the Earth and some smaller. They are much further away than the Moon.

Like the Moon, the planets shine only because they are being lit up by the Sun. When you look at them they seem like stars In the sky, but to see them well you must use a telescope.

Opposite: Mars as seen from space. *NASA*

MERCURY

The surface of Mercury, the planet closest to our Sun. *Paul Doherty.*

VENUS

The brightest planet is Venus. It is nearly as big as the Earth, but it is closer to the Sun than we are, and is so hot that we could not live there.

Volcanoes on Venus. *Paul Doherty.*

Venus, where only part of its surface is in sunlight. *Paul Abel.*

MARS

This is Mars as it usually looks. *NASA*

The planet Mars is red, smaller than the Earth, and further from the Sun. It has very thin air, which we could not breathe, and is very cold. The poles are covered with ice, just like our north and south poles.

Here, a dust storm hides the surface of the planet. *NASA*

Jupiter from one of its moons. *Tony Wilmot.*

JUPITER

The biggest planet is Jupiter. You can see it for several months in the year. It has a surface made of gas, so you could not go there. Telescopes show that on it there is a big patch called the Great Red Spot. You will also see what look like four small stars close to Jupiter. These are Jupiter's moons.

The giant planet Jupiter, taken by NASA's Hubble Space Telescope. The black spot on the planet is the shadow of one of its moons, Io.

SATURN

The other big planet is Saturn. It looks like a bright star. Like Jupiter, it has a surface made of gas. With a telescope you can see that it has rings. The rings are made up of small bits of ice moving round Saturn, and there is one big moon, called Titan. Rockets have been sent to Jupiter and Saturn.

The planet Saturn with its rings, photographed with my 15-inch telescope. *Damian Peach.*

Saturn in 2009, when the rings were almost edge-on. The white spot to the left is Reah, one of Saturn's moons. *Paul Abel.*

Saturn as seen from its moon, Titan. *Tony Wilmot.*

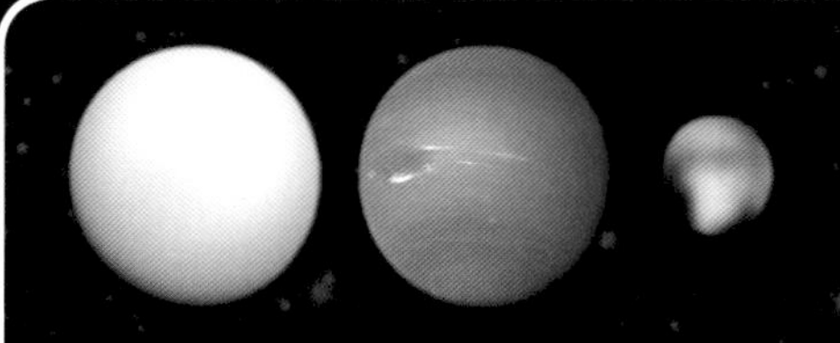

Further away from Saturn there are two more large planets, Uranus and Neptune, and many hundreds of smaller bodies, of which the brightest is called Pluto. Pluto's 'year' is 248 times as long as ours, because it moves much more slowly around the Sun and is so far away.

CHAPTER SIX

Comets and Shooting Stars

Sometimes you can see a streak of light in the night sky. We call this a shooting star, but it is nothing to do with a real star. It is a tiny bit of dust, burning up in the air at more than forty miles above you.

August is the best time to look for shooting stars. Astronomers call them 'meteors'.

Opposite: The brilliant Donati's Comet, seen in 1858.

A green comet called Lunin. *Greg Parker.*

Above: The Great Comet of 1843, bright enough to cast shadows. *Paul Doherty.*

Left: A drawing of the comet, as seen in 1811. *Paul Doherty.*

Above: A spaceship inside Halley's Comet.

Below: A falling meteorite.

Meteorites

Sometimes a bigger stone or rock falls from the sky. Astronomers call them 'meteorites'. They may make craters like those of the Moon. Craters are not made often, and we do not know of anyone that has been killed by a meteorite. The picture above is of a meteorite that had been moving around the Sun for millions of years.

Jessica Sutton holding the Barwell Meteorite, which is 4.6 billion years old.

The meteor that is said to have killed all the dinosaurs 65 million years ago. *Paul Doherty.*

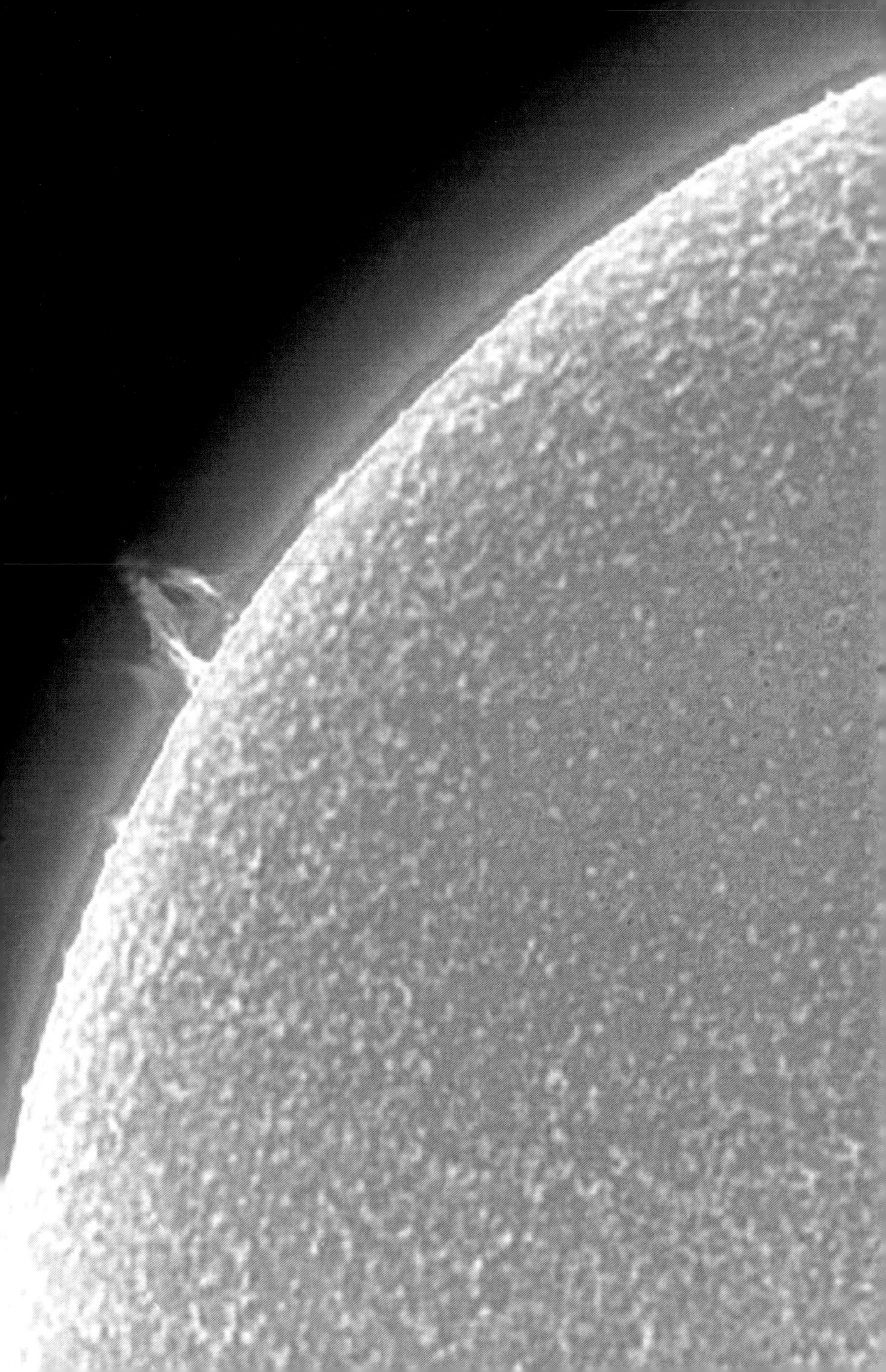

CHAPTER SEVEN

Our Star: The Sun

Above: Sunspots

To see the Sun well, we can use a telescope to show it on a screen. Sometimes there are dark patches called sunspots. They are not really black, but are less bright than the surface around them.

Because the Sun's surface is made of gas, no sunspot can last for very long. The Sun spins round, taking several weeks to make one turn, so that the spots seem to move from day to day. A big spot may be much bigger than the Earth, but sometimes there are no spots at all.

Opposite: Masses of gas rising from the Sun's surface. *Alan Clitherow.*

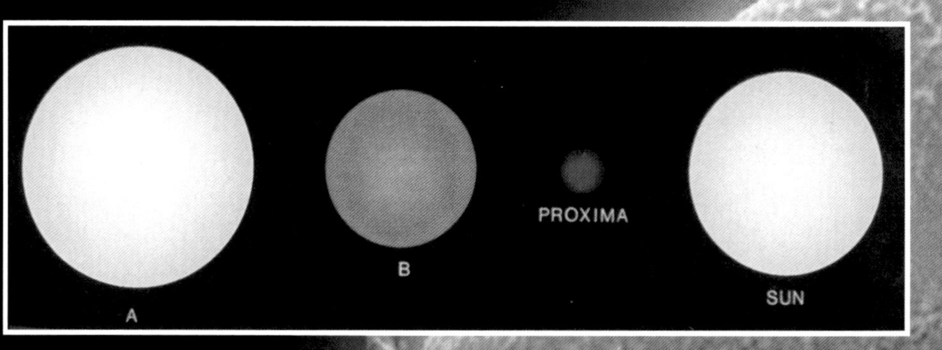

The size of the Sun compared with the nearest stars.

Eclipses

The Earth goes round the Sun, and the Moon goes round the Earth. Sometimes the Moon moves between the Earth and the Sun, and covers it for a few minutes. This is called an eclipse of the Sun. If all of the Sun is hidden we see the 'corona' gas round the Sun. Full eclipses are lovely; from England there will not be another until 2087, but there will be some in other parts of the world.

A photo of the Great Asian Solar Eclipse on 22 July 2009.

There are also eclipses of the Moon, but these are different — the Moon is closer than any other body, so nothing can get between it and the Earth. But the Moon may move into the Earth's shadow; the sunlight cannot fall on to it, so that the Moon becomes faint until it moves out of the shadow. Eclipses of the Moon happen quite often.

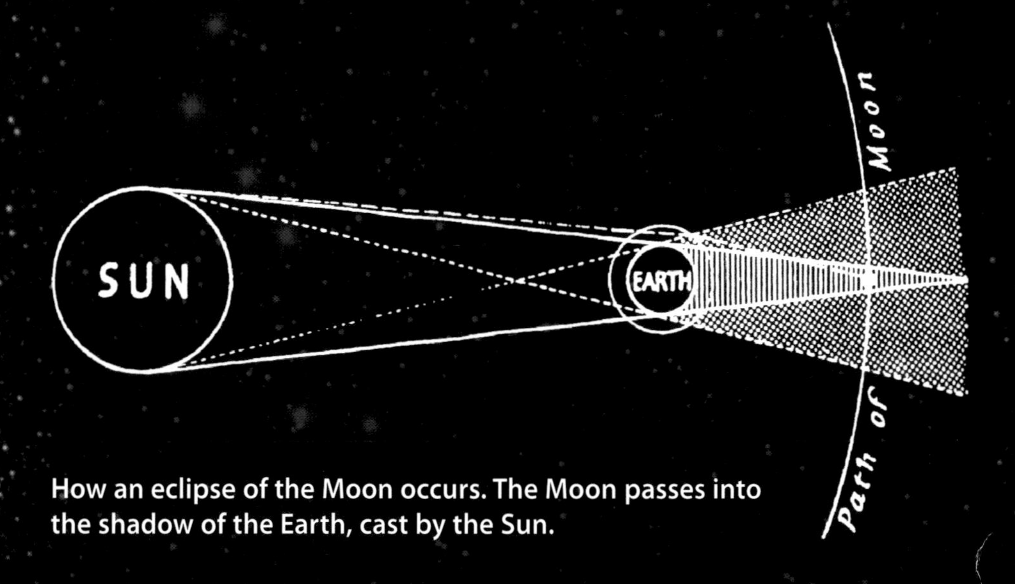

How an eclipse of the Moon occurs. The Moon passes into the shadow of the Earth, cast by the Sun.

An eclipse of the Moon. The Earth's shadow passes across the Moon. *John Fletcher.*

CHAPTER EIGHT

The Spinning Sky

The stars are suns, but they are so far away that telescopes show them only as points of light. If we wanted to make a model, and put the Earth and the Sun one inch apart, even the nearest star would have to be put over four miles away! This is why the stars look so much fainter than the Sun.

Opposite: A star field. Each of these stars is a sun. *Greg Parker.*

The Earth spins round once in twenty-four hours. This means that the whole of the sky seems to move round us once in twenty-four hours, taking the Sun, Moon, stars and planets with it. The Earth spins from west to east, so that the sky seems to spin the other way. This is why the Sun rises in the east and sets in the west.

This photo shows a line of stars in the sky. The stars are not really the same distance from us. *Greg Parker.*

CHAPTER NINE

The Constellations

Because the stars are so far away, we do not see them moving compared with each other. They are spread out in what are called 'constellations', and these constellations do not change. The Sun, Moon and planets are much closer, so that they move slowly around from one constellation to another.

Opposite: A view from a planet inside the star cluster M13. *Paul Doherty.*

Left: Centaurus A — two star systems in space, one lies in front of the other. *Paul Doherty.*

Below: A system of planets in a star cluster a long way away. *Paul Doherty.*

Penguins looking up at the Southern Lights. The Southern Lights are a natural light display in the sky and can be seen at night near the South Pole. *Paul Doherty.*

Rings around a star called Vega. *NASA*

People who lived thousands of years ago made up stories about the constellations, and we still use the old names. The best known of our constellations is the Great Bear because of the way in which its seven main stars are placed. Many people call it the Plough, and Americans call it the Dipper. From England you can see it on any clear night.

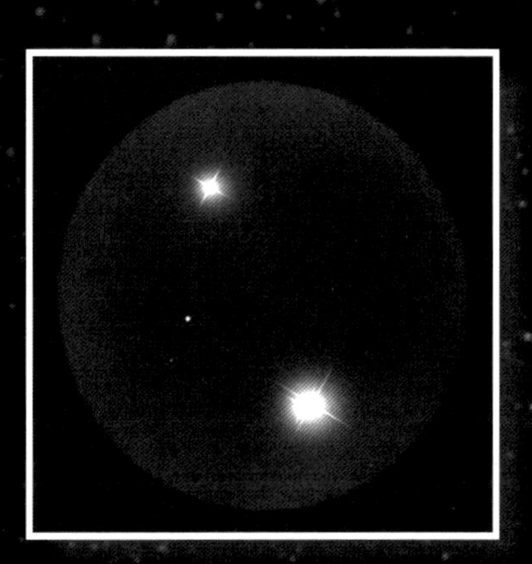

Right: The double star, Mizar, in the Great Bear. *John Fletcher.*

Below: The Great Bear, the Little Bear and the Pole Star.

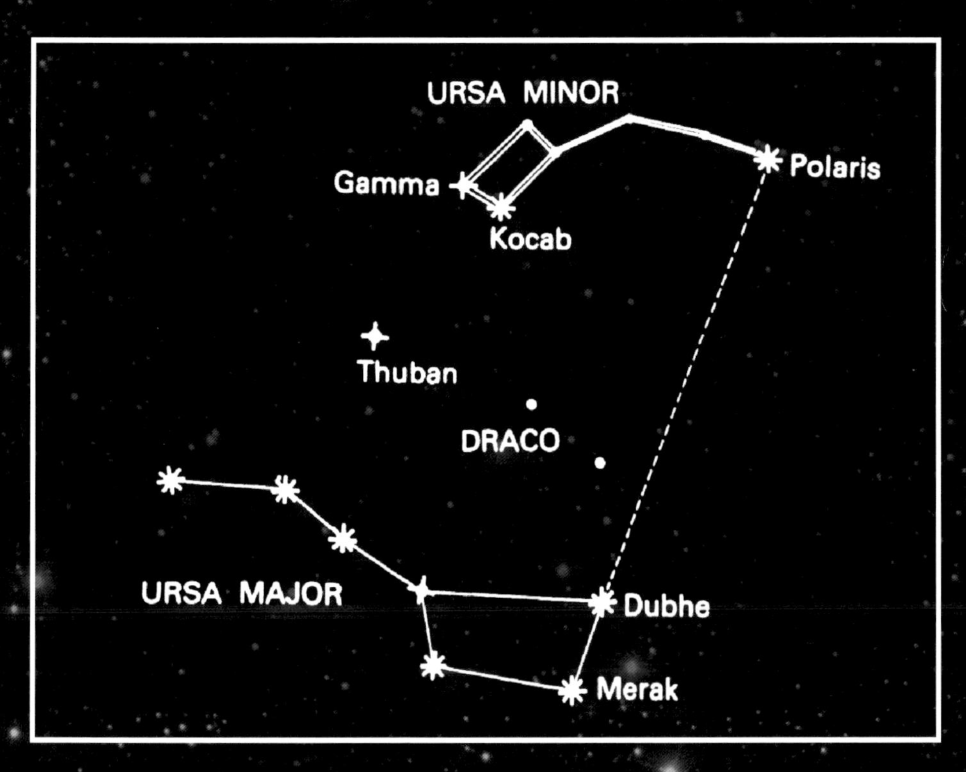

The scene from a planet moving around the star Castor. *Paul Doherty.*

The sky from a planet going around the star Mizar. *Paul Doherty.*

Orion: The Hunter

Another famous constellation is Orion, the Hunter. Its stars are brighter than those of the Great Bear, but you cannot always see it; look for it on winter nights, but in summer it is too near the Sun in the sky. It is not easy to see a hunter in the pattern of stars in Orion!

The stars are not all at the same distance from us, so that the stars in a constellation are not really close to each other. In Orion, there are two very bright stars, named Betelgeux (in the Hunter's shoulder) and Rigel (in his foot). They do not look far apart in the sky, but Rigel is twice as far away as Betelgeux. The diagram above shows what is meant. (The star names are very old. Betelgeux is often called 'Beetle-juice'!)

CHAPTER TEN

The Colours of Stars

The stars are not burning in the same way as a coal fire, but they are very hot, and they last for a very long time. When you were born our Sun looked the same as it does now, and you will still look the same when you have grown old, but it will not go on shining for ever.

The stars are not all the same colour. Our Sun is yellow; some stars are white, some are blue and some are red. This means that some stars are hotter than others. Blue and white stars are hotter than our yellow Sun; red stars are cooler. In Orion, Betelgeux is red and Rigel is white. Rigel shines as brightly as 40,000 suns put together!

Opposite: A very big red star with a smaller blue companion. *Paul Doherty.*

The life of a star, from a red giant to a white dwarf.

An exploding star

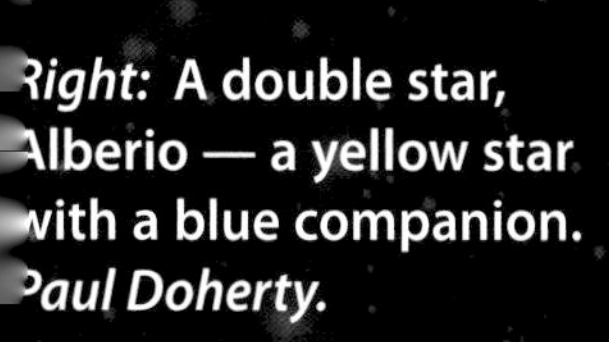

Right: A double star, Alberio — a yellow star with a blue companion. *Paul Doherty.*

Below: The planet of a giant red star. *Paul Doherty.*

CHAPTER ELEVEN

The Milky Way

Go out on a clear, dark night, and you will see a belt of light crossing the sky. This is called the Milky Way, and if you look carefully you will see that it is made up of stars — so many stars you cannot hope to count them. They seem to be so close together that they almost touch each other, but they are really a long way apart.

Opposite: Glowing gas in the Milky Way. *Paul Such.*

Our Sun is one of the stars in what we call the Galaxy. The Galaxy is shaped rather like the wheel of your bicycle — flat, with a bump in the middle. When you look along the wheel-shaped Galaxy, as the boy in the picture is doing, you will see many stars, almost one behind the other. This gives the appearance of the Milky Way.

This picture shows the sky from Virginia, USA with the Milky Way to the left and the planet Jupiter to the right. *Kent Blackwell.*

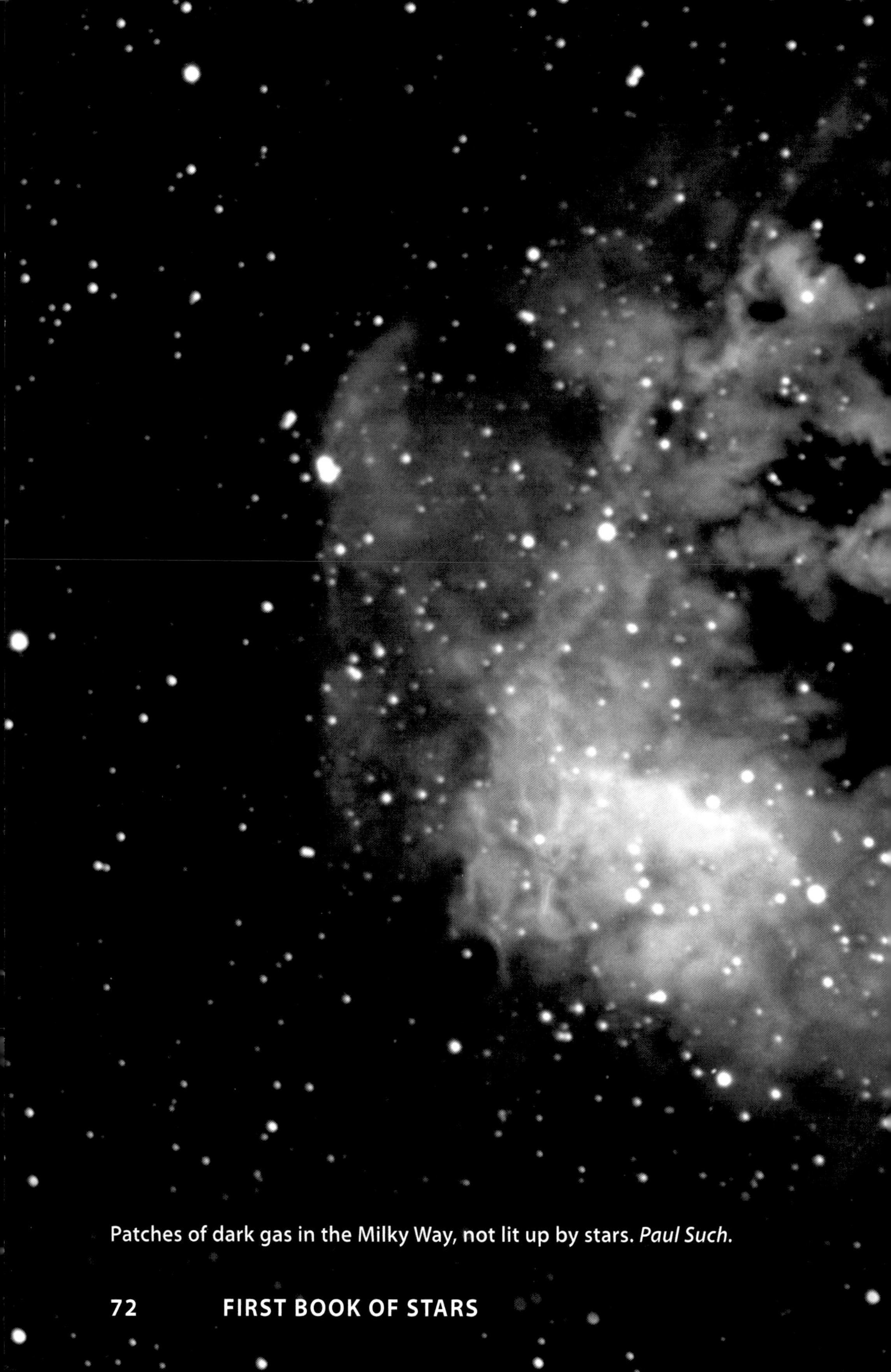

Patches of dark gas in the Milky Way, not lit up by stars. *Paul Such*.

CHAPTER TWELVE

Galaxies

We know that our Sun is one of the stars in our Galaxy. There are many other galaxies, a very long way away indeed. Some of them are 'spiral', as shown here. We have found that even our own Galaxy is spiral. The Sun, with the Earth and the other planets, is just outside one of the 'arms' of the spiral, well away from the middle of the Galaxy.

The galaxies are so far away that they look faint, and to see them you must use a telescope. We do not know how many galaxies there are; the Universe is very large, and the Earth is not important except to you and me!

Opposite: A spiral galaxy as seen by the Hubble Space Telescope. *NASA*

A cluster of stars. If we lived within the cluster the sky would have many brilliant stars shining at night. *Richie Jarvis.*

The Andromeda galaxy, a spiral about the same size as our own galaxy, so far away that the light takes 2 million years to reach us. *Richie Jarvis.*

M33, a spiral galaxy smaller than ours, that lies in a constellation called Triangulum. *Nick Howes.*

Two distant systems of stars shaped like spirals. *Alan Clitherow.*

The Whirlpool Galaxy, a lovely spiral galaxy. *Nick Szymanek.*

A cluster of galaxies. Each of these patches contains millions of stars.

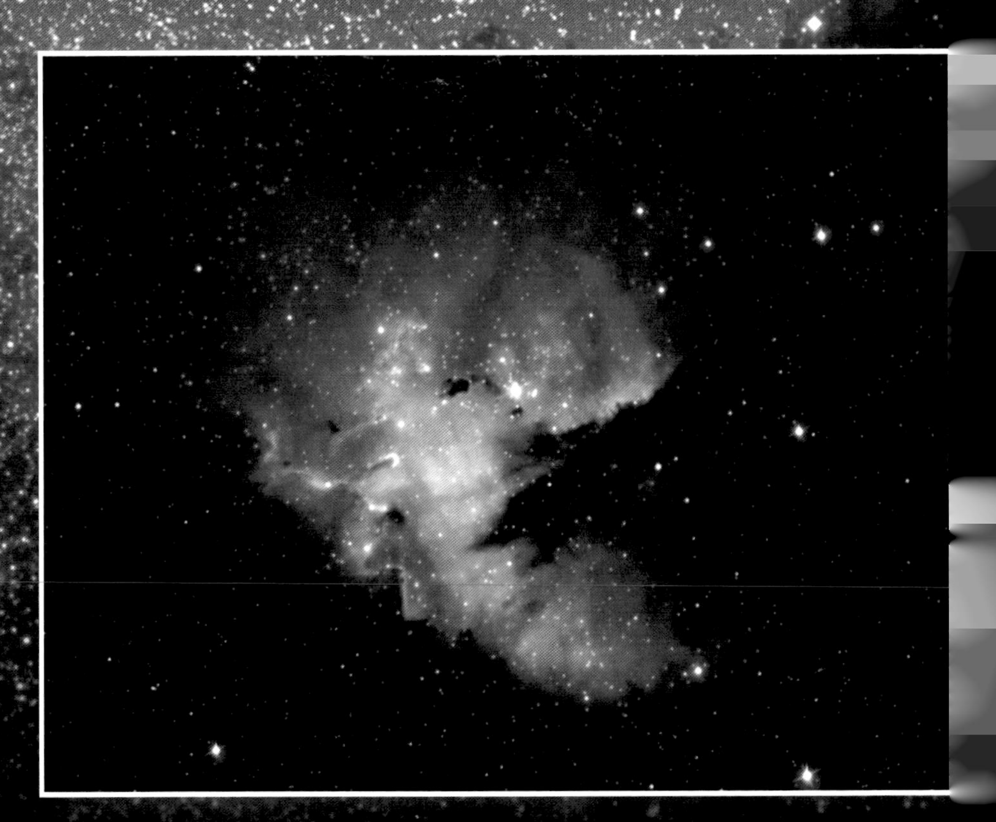

A nebula — a patch of gas in the sky. *Peter Such.*

Nebulae

There are also patches, which we call 'nebulae', which is the Latin word for 'clouds'. They really are clouds of gas. Inside these nebulae new stars are being born. There is one famous nebula in Orion; we call it the Hunter's Sword. Long ago, our Sun was born inside a nebula of this kind.

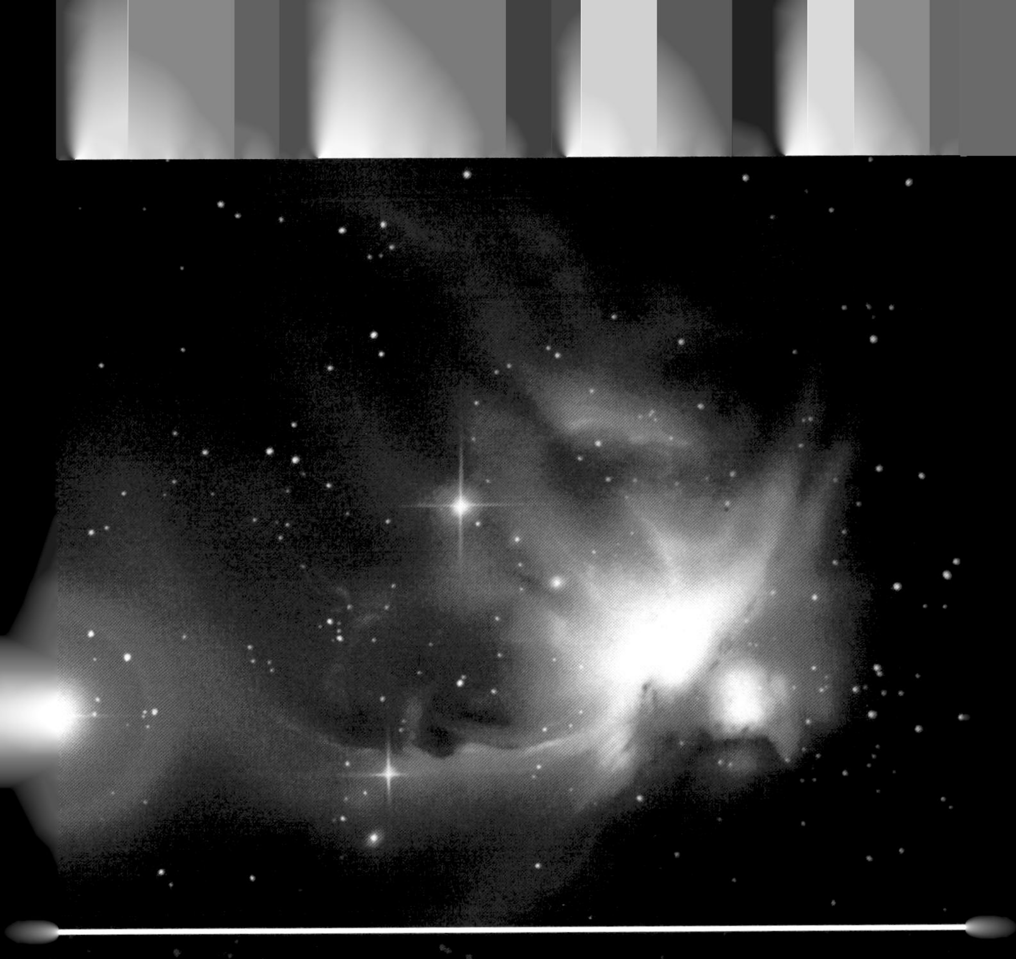

The Orion Nebula. *Peter Such.*

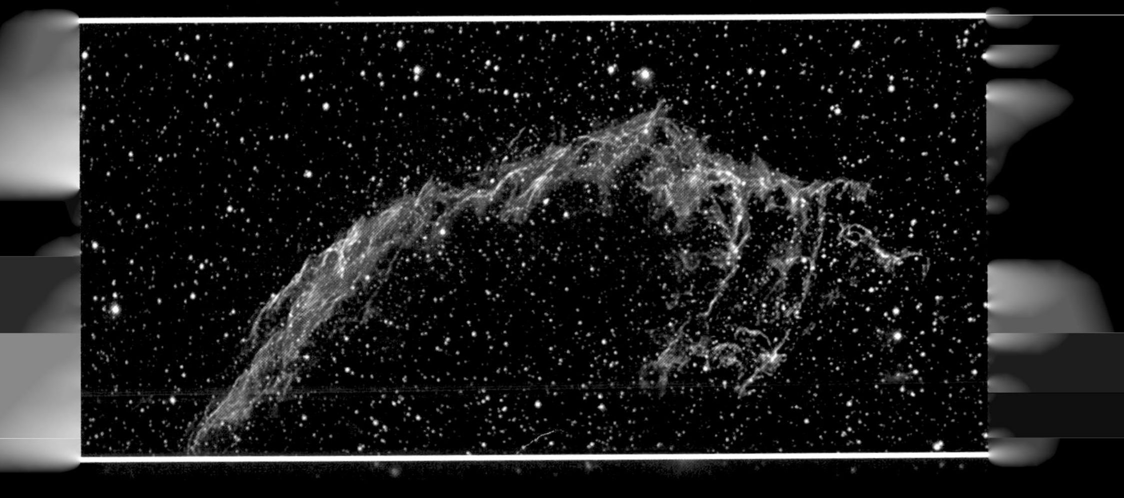

The California Nebula, 1140 light years away. *Gordon Rogers*.

The North American Nebula, a star system that has the shape of America

Nearing a nebula. *Paul Doherty*.

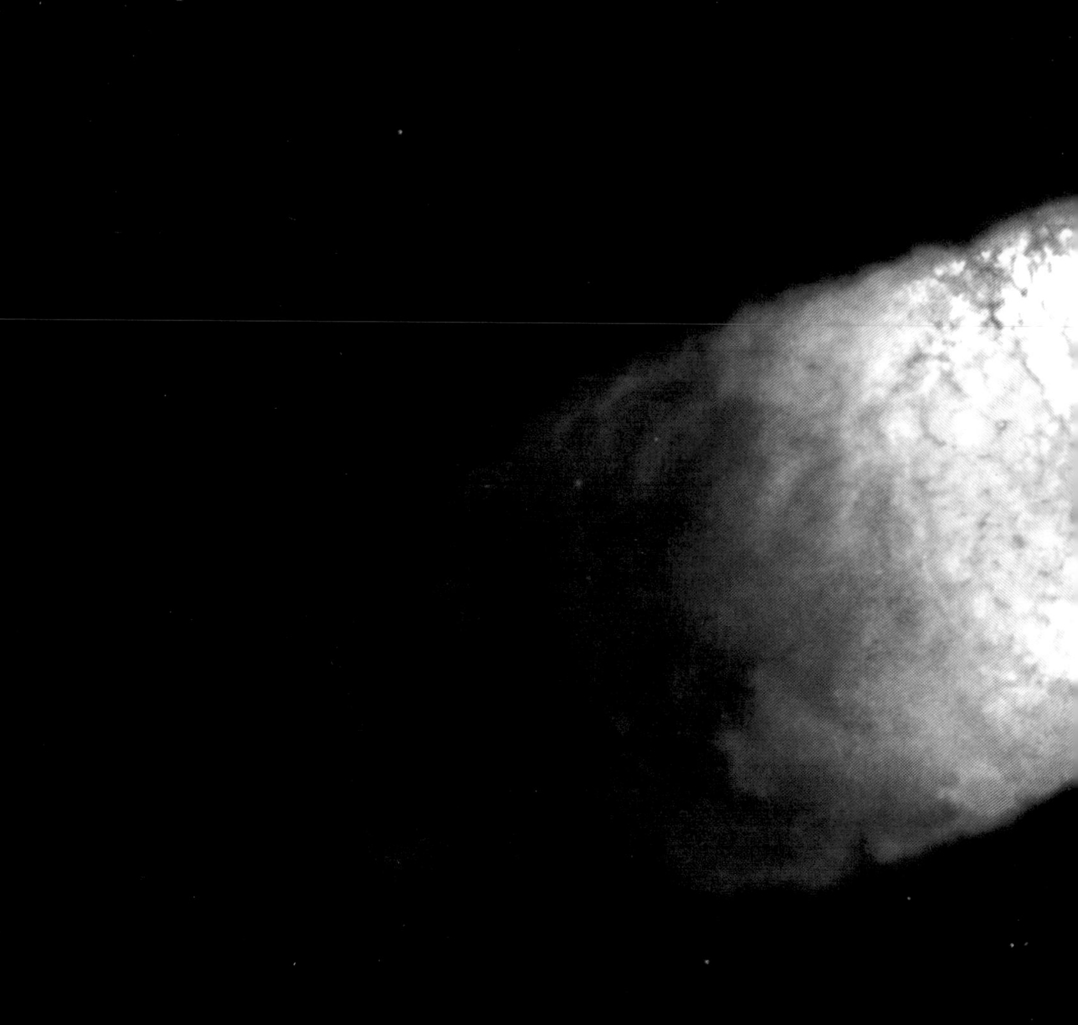

A funny looking planetary nebula, an old star that has thrown away its outer layers. *NASA*

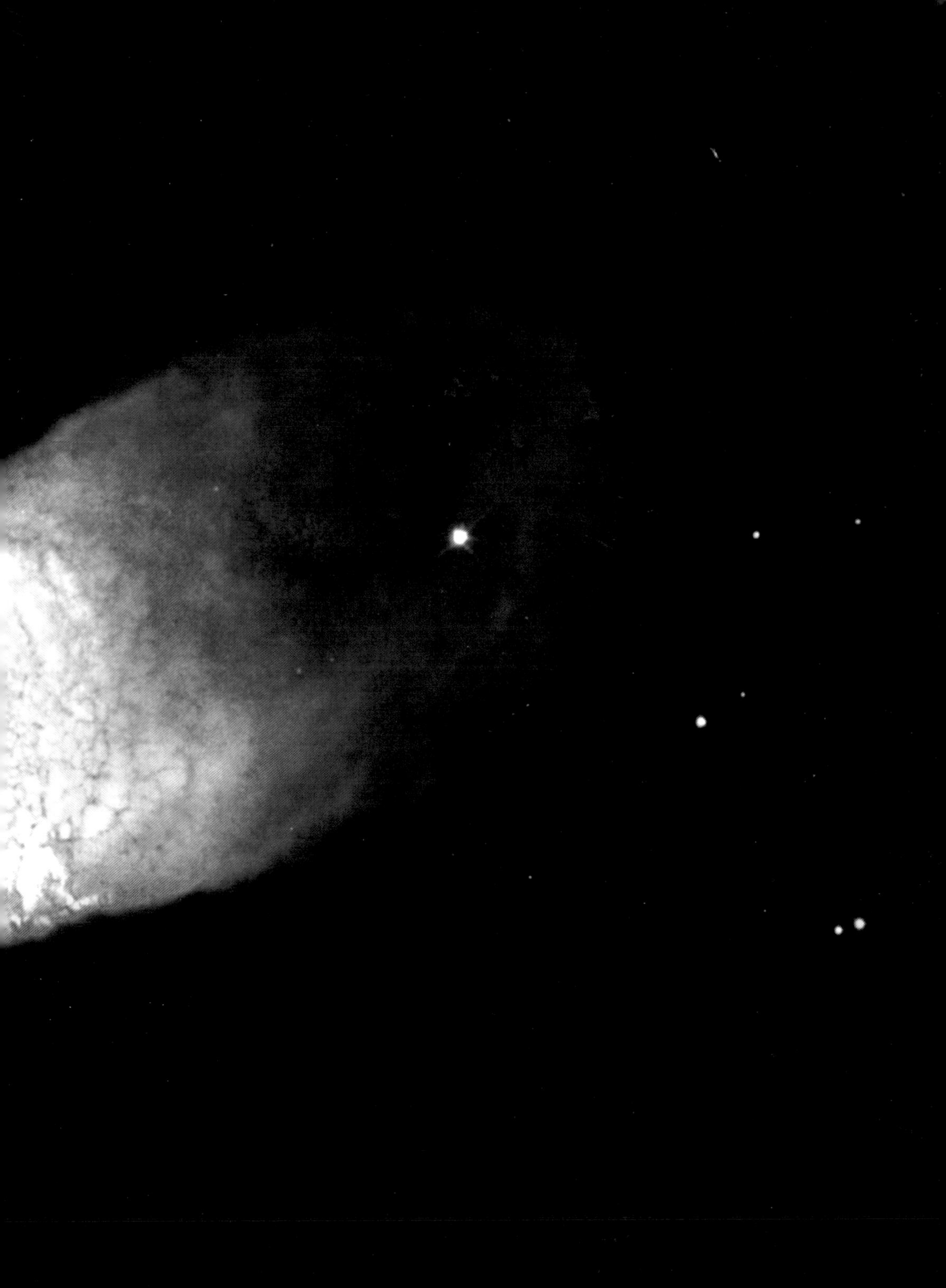

The Crab Nebula, all that remains of a star that was seen to blow up over 1000 years ago. *Richie Jarvis.*

A large black hole. *Paul Doherty.*

Black Holes

When a large star runs out of fuel it can no longer support its heavy weight. The pressure from the star's massive layers of hydrogen press down forcing the star to get smaller and smaller and smaller. Eventually the star will be even smaller than an atom. The force of gravity in the star is concentrated and magnified, becoming a black hole.

A black hole is a region of space from which nothing, not even light, can escape.

CHAPTER THIRTEEN

Life Beyond the Earth?

Are there men and women on other worlds? There seems no reason why not, but we cannot be sure. We know that there are no people on the Moon or any of the planets that move round the Sun, but other stars also have planets and there may be planets just like the Earth.

We cannot hope to send rockets to planets of other stars. They are much too far away, so that if people live there we cannot visit them. The best chance of getting in touch with them would be by sending radio messages, but we do not yet know if they really exist — and even if so, whether they are at all like us.

Opposite: Are dragons living on other planets far away? *Paul Doherty.*

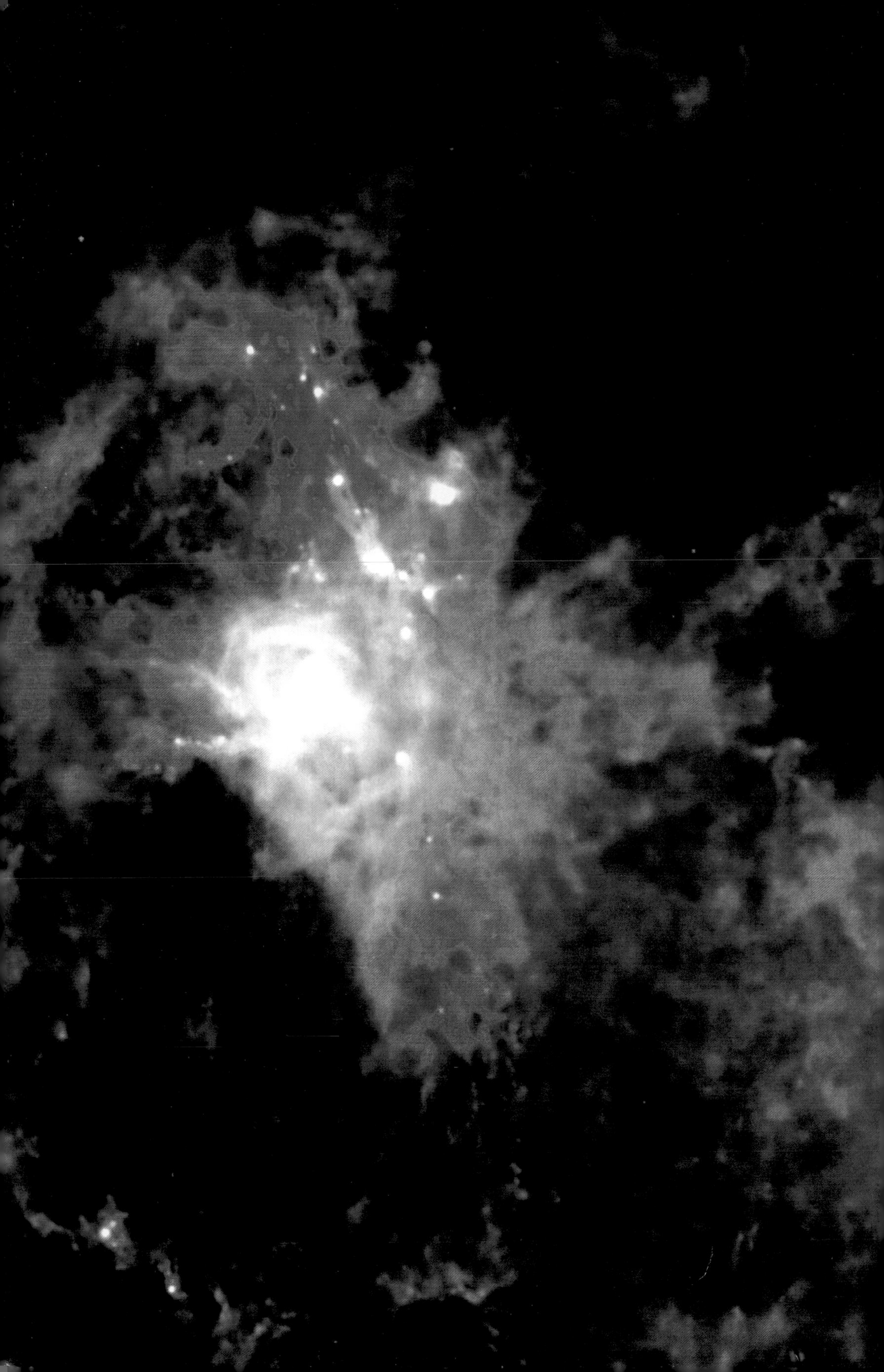

Some Helpful Notes

Let us end by looking back at some of the things I have told you in this book.

The **EARTH** is a planet, moving round the Sun. The **SUN** is a state made of gas. It seems so much brighter and hotter than the other stars because it is so much closer to us. NEVER LOOK RIGHT AT THE SUN FOR LONG; you would hurt your eyes.

The **PLANETS** have no light of their own. They shine only because they are being lit up by the Sun. The **MOON** moves round the Earth. It also shines because it is lit up by the Sun.

A **SHOOTING STAR** (a meteor) is a tiny piece of rock or dust burning away in the upper air. It is quite unlike a real star. All **STARS** are Suns, many of them much bigger and hotter than our Sun. The star patterns in the sky are called **CONSTELLATIONS**. They have names, but the stars in a constellation are not really close to each other.

A **NEBULA** is a gas-cloud in which new stars are being born. A **GALAXY** is made up of a very great number of stars. The Sun is just one star in our own Galaxy. The **UNIVERSE** contains all the galaxies — we cannot really appreciate how big it is!

What Next?

Have you enjoyed this book? If so, you may want to find out more. This is what I think you should do...

Read some more books, more detailed than mine.

Find someone who knows about the sky, and see if he or she will help.

See if there are any astronomical societies that you can join.

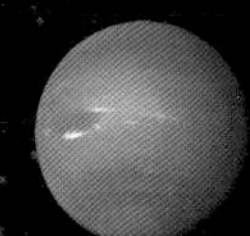

Watch any astronomical programmes on television (*The Sky at Night* is shown every month).

And most important of all — go outdoors on a clear, dark night and learn your way around the sky.

The best of luck to you!
Sir Patrick Moore